Animal Needs:
Who's New at the Zoo?

by Emily Sohn and Barbara J. Foster

Chief Content Consultant
Edward Rock
Associate Executive Director, National Science Teachers Association

NorwoodHouse Press
Chicago, IL

Norwood House Press
PO Box 316598
Chicago, IL 60631

For information regarding Norwood House Press, please visit our website at
www.norwoodhousepress.com or call 866-565-2900.

Special thanks to: Amanda Jones, Amy Karasick, Alanna Mertens, Terrence Young, Jr.

Editors: Michelle Parsons, Diane Hinckley
Designer: Daniel M. Greene
Production Management: Victory Productions, Inc.

Paperback ISBN: 978-1-60357-279-8

The Library of Congress has cataloged the original hardcover edition with the following
call number: 2010044547

Manufactured in the United States of America in North Mankato, Minnesota.
296R—082016

Contents

Note to Caregivers:

Throughout this book, many questions are posed to the reader. Some are open-ended and ask what the reader thinks. Discuss these questions with your child and guide him or her in thinking through the possible answers and outcomes. There are also questions posed which have a specific answer. Encourage your child to read through the text to determine the correct answer. Most importantly, encourage answers grounded in reality while also allowing imaginations to soar. Information to help support you as you share the book with your child is provided in the back in the **Additional Notes** section.

Words that are **bolded** are defined in the glossary in the back of the book.

What Makes an Animal an Animal?

Quick: Name as many animals as you can. How many did you come up with? Chances are, you missed a lot. There are thousands of types of animals on Earth.

In this book, you will learn about what makes animals the same. You will also learn what makes some types of animals unlike others.

As you read, you will solve a puzzle. What animal will live in the new **exhibit** at the zoo?

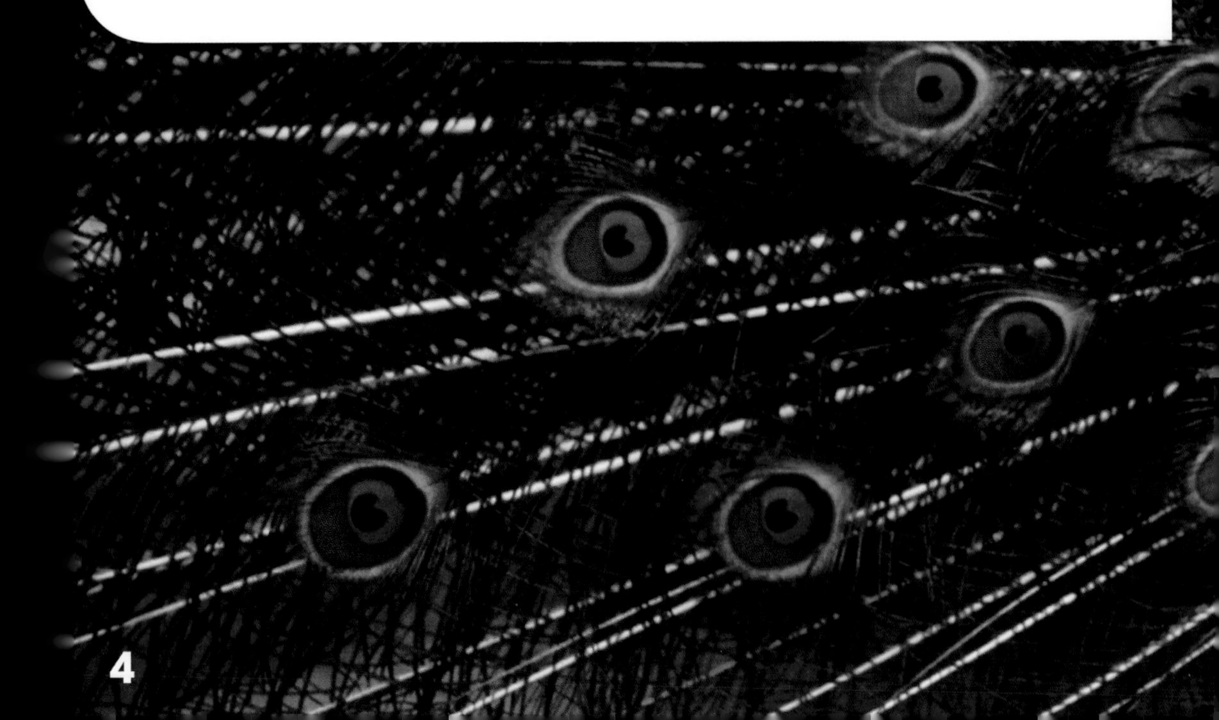

Which Animal Lives Here?

You visit the zoo. There is a new exhibit. It is the size of the end zone on a football field, which is 53 yards (48 meters) wide and 10 yards (9 meters) deep. There are pools of water. The water is three or four feet (about 1 meter) deep.

You also see lots of rocks. There are no trees. There is some ice floating in the water. But the animals are missing. Maybe they are sleeping. Or maybe they escaped!

Animal 1: Clown Fish

Animal 2: Emperor Penguins

Can you guess which animals belong in the exhibit?

Animal 3: Box Turtles

Animal 4: Seagulls

To solve the puzzle, think about these questions:

1. What does each animal eat?

2. How does each animal move?

3. What kind of weather does each animal like?

4. How much space does each animal need?

Guess the Animal

Grab a friend and play a game. Tell your friend to think of an animal. Try to guess the animal. To do this, ask questions. Your friend will give only "Yes" or "No" answers. For example, you could ask: "Does it live in water?" Try to ask about things the animal needs to survive.

How is a squirrel different from a pigeon?

How many questions did it take to guess the answer? Which questions worked best?

What Is an Animal?

What is your favorite animal? Ask your friends about theirs. They may say dogs or dolphins or sponges or snakes. It can be hard to see how all of these creatures are alike. But there are things that all animals share in common.

For one thing, all animals are alive. That sets them apart from nonliving things, like rocks. All animals eat food. And all animals move at some points in their lives.

Other living things include **plants, algae,** and **fungi.**

Look at the creatures that might be new at the zoo. What do they share in common? How are they different from plants?

The Six Main Groups of Animals

Animals are like sports players. They all belong to some kind of group, like a club or a team. There are six main groups of animals.

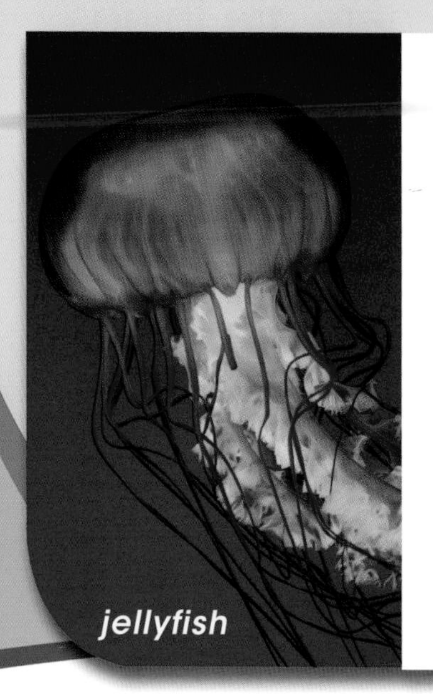
jellyfish

tiger

hummingbird

1. **Invertebrates** do not have a backbone. This group includes worms, octopuses, and jellyfish.
2. **Mammals** have hair or fur. People (That's right! You're an animal, too!), tigers, and seals belong in this group.
3. **Birds** have feathers and a **bill.** Hummingbirds, penguins, and seagulls fit into this group.

4. **Amphibians** breathe with **gills** in early life. As adults, they have **lungs.** The group includes salamanders, toads, and frogs.
5. **Reptiles** have scaly skin and creep on the ground. Some can climb, too. Snakes, turtles, and lizards fit into this group.
6. **Fish** have fins and gills. Sharks and clown fish belong to this group.

Look at the zoo puzzle choices. To which groups do they belong?

How Do Animals Live?

Every animal has to eat. Some eat other animals. For example, polar bears eat seals. Some animals eat just plants. Bison eat grass. Some eat both animals and plants.

Zookeepers feed zoo animals every day. At the zoo, animals get food that is like what they eat in the wild.

No Place Like Home

Like you, all animals have a home to live and sleep in. Fish live in water. Deer live on land. Penguins eat in water. But they lay eggs on land.

Do you like it cold? Then, you are like an Emperor penguin. Where they live, there is snow and ice all year long. Salmon and cod also live in cold waters. But the water they live in is not as cold as the Emperor penguins' water.

giraffe

deer

Do you like it hot? Then, you are like a giraffe. They live where leaves can grow on trees all year long. Clown fish live in warm waters.

Think about the zoo puzzle. There is ice in the exhibit. So it must be cold. Which animals in the zoo puzzle would be comfortable there?

On the Move

Zoo animals need space to move. There are many ways that creatures get around. Seagulls are birds that can fly. Most birds fly, but not penguins. Their wings are too small. Instead, they swim, waddle, and slide on their stomachs on ice. Polar bears and turtles also swim, and they walk on four legs. Snakes and worms slither.

Can you think of other ways that animals move? How does a **centipede** move? How does a monkey move? How about a frog? How many ways can you move?

Which animals in the zoo puzzle do you think would be able to move well in the empty zoo exhibit?

How Much Space Do Animals Need?

Many kids like big spaces. On a field or playground, there is lots of room to run. Animals also need a certain amount of space. Some animals need just a little room. A clown fish can live near a small area of coral.

Whales need large areas of ocean. Lions need lots of land. Seagulls need room to be able to fly.

Can you think of other animals that need a little space? What animals need a lot of space?

Look at the zoo puzzle choices. How much space does each animal need?

What Can Be Bad for Animals?

Life can be hard for animals in the wild. They have to find food. They have to escape other animals that want to eat them. If they get hurt, there is no one around to help. Wild animals can also get sick.

oil-covered scallops

cheetah looking for food

People make life even harder for wild animals. We build cities and homes. That destroys their homes. Animals get sick from our trash and chemicals. Some factories dump chemicals into rivers. Sometimes we spill oil, which gets into the soil, lakes, and rivers. Animals have no way of fighting back.

Raccoons can get rabies.

Some animals can get the flu. Some can get rabies. And they can get sick in other ways. Sick animals do not always look sick. You should stay away from all wild animals. Some sick animals can make people sick, too.

How might zoos keep animals healthy? Why might experts want to study sick animals?

Science at Work

Veterinarian

Veterinarians are animal doctors. They are often called vets.

Vets can give out medicines. Some help sick animals get better. Others keep animals from getting sick in the first place.

Vets can fix a bird's broken wing, a turtle's broken shell, and a cat's broken leg. Some vets work on horses, bears, and giraffes!

Vets need to know what an animal eats. They need to know where it likes to live. And they need to know what it needs to be healthy.

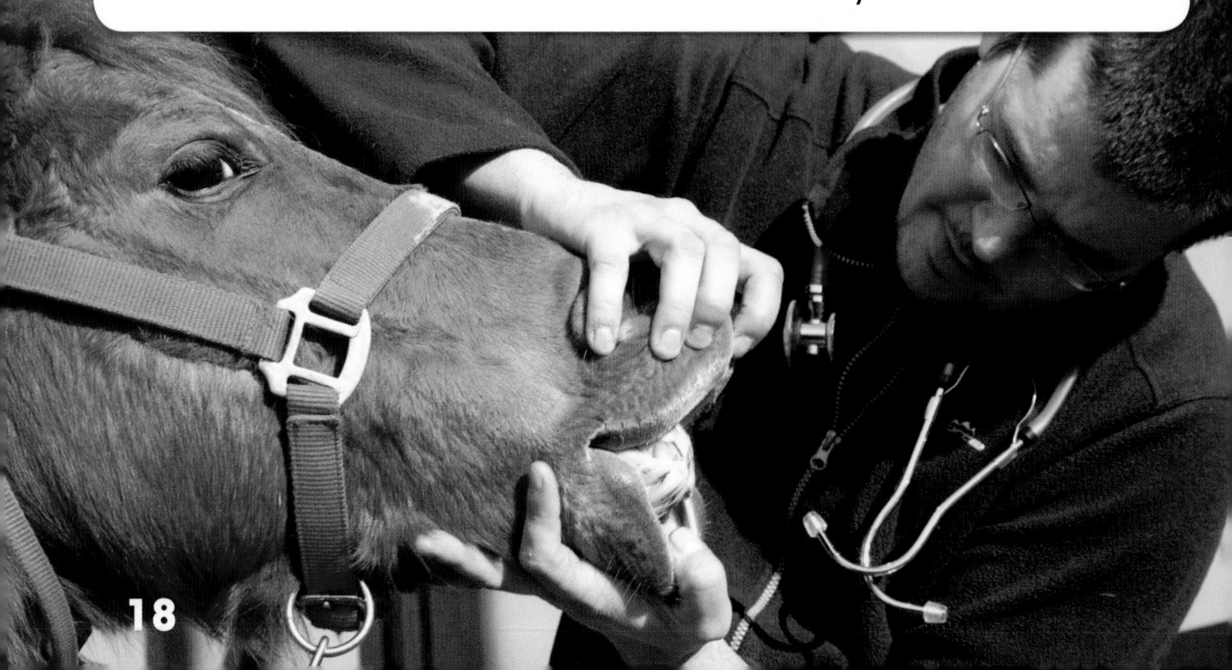

This statue in Central Park in New York City honors Balto, the dog hero.

Connecting to History

Balto, Dog Hero

It was 1925. Kids were getting sick in Nome. That's a city in Alaska. The only medicine was in Anchorage. The cities are nearly 1,000 miles (1,609 kilometers) apart. It was too cold to start a plane. So, the medicine started on a train. Then, dogs pulled it to Nome on a sled. Different dog teams worked at different stages of the trip. The lead dog of the final team was named Balto. He became a hero.

Solve the iScience Puzzle

Let's return to the empty zoo exhibit. To solve the puzzle, think about what each animal needs.

Animal 1: Clown fish live in warm water. They eat tiny animals and algae. They need only a little space.

Animal 2: Emperor penguins live in cold water and on cold land. They eat fish. They need room to waddle, slide, and swim.

Animal 3: Box turtles live near trees and other plants. They eat insects, mushrooms, and fruits. They don't need much space.

Animal 4: Seagulls usually live near water. But they can live away from water if there is food to eat. They eat fish, insects, clams—and edible waste that people drop.

Which animal do you think belongs in the zoo exhibit? Why?

Beyond the Puzzle

You figured out which type of animal lives in the new exhibit. The zoo home is made for that animal. Now, how would you decide whether to have a dog, a parrot, or a goldfish as a pet? Which animal would be best for your home?

You would have to think about each animal's needs. You would have to choose the pet you could best take care of.

Glossary

algae: living things that make their own food, as plants do, but that have no roots or leaves.

bill: a bird's beak.

centipede: animals that have between 15 and about 177 sections, each with a pair of legs.

exhibit: something set up to show to people who come to see it.

fungi: organisms such as mushrooms, yeasts, and molds that feed on organic material.

gills: body parts that let an animal breathe underwater.

lungs: body parts that let an animal breathe in air.

plants: living things that make their own food and cannot move from place to place.

veterinarians: doctors who take care of animals.

Further Reading

The ABCs of Habitats, by Bobbie Kalman. Crabtree Publishing Company, 2007.

Animals at Home, by David Lock. DK Readers, 2007.

I See A Kookaburra: Discovering Animal Habitats Around the World, by Steve Jenkins and Robin Page. Houghton Mifflin Company, 2005.

Animal Homes Games.
http://www.kidport.com/grade1/science/animalhomes.htm

Animal Homes Activities.
http://www.nwf.org/Get-Outside/Be-Out-There/Activities/Observe-and-Explore/Search-for-Animal-Homes.aspx

Additional Notes

The page references below provide answers to questions asked throughout the book. Questions whose answers will vary are not addressed.

Page 9: Animals cannot make their own food. Plants can make their own food.

Page 11: Clown fish is a fish. Emperor penguin is a bird. Box turtle is a reptile. Seagull is a bird.

Page 14: A centipede walks on its many pairs of legs. Some kinds of monkeys swing through the trees, using their long arms and tails. Some kinds of monkeys walk along the ground on their front and back paws. Frogs hop.

Page 17: Zoos can keep animals healthy by making sure the animals have the right food and that they are warm or cool enough. People who work at the zoos can check the animals to make sure they are well, and take care of them when they are sick. Experts might study sick animals to find out what makes them sick and how they can get better.

Index